MW00915145

THOMAS EDISON

THE ONE WHO CHANGED THE WORLD

THE HISTORY HOUR

Copyright © 2018 by Kolme Korkeudet Oy

All rights reserved.

No part of this book may be reproduced in any form or by any electronic or
mechanical means, including information storage and retrieval systems,
without written permission from the author, except for the use of brief
quotations in a book review.

CONTENTS

PART I
Introduction   1

PART II
BORN INTO A COUNTRY OF GREAT CHANGE
Edison's disability   5
A different kind of education   7

PART III
THE TELEGRAPH AND THE START OF HIS
GENIUS
Edison's curious mind   11
His path to stardom   13

PART IV
HOW EDISON CREATED THE INVENTION
MACHINE
The Wizard of Menlo Park   17
The home of the light bulb   19

PART V
EDISON AND THE PHONOGRAPH
Other notable inventions   27

PART VI
IT WASN'T ALWAYS ABOUT SUCCESS FOR
EDISON
The biggest failure of all   35

PART VII
THE TRUTH ABOUT THE LIGHT BULB
The real inventors   39
Bamboo was the answer   41

PART VIII
THE RIVALRY BETWEEN TESLA AND THE "WAR
OF THE CURRENTS"

Edison's gruesome experiments   47

PART IX
Father of the motion pictures   51

PART X
THE MAN BEHIND THE INVENTIONS

He lived for his work   57

Edison kept great company   59

PART XI
EDISON'S GREATEST QUOTES

The timeline of Edison's life   69

PART XII
HIS DEATH AND THE LEGACY HE LEFT BEHIND

The legacy of Edison   73

The strengths and weaknesses of Thomas Edison   76

How can we use Edison's strengths in our lives?   79

PART XIII
REMEMBER HIM FOR THE RIGHT REASONS

The best books on Thomas Edison   85

Your Free eBook!   87

INTRODUCTION

❦

Thomas Edison, the inventor of the light bulb. Except there are a lot more to the story than just the light bulb, and there's a lot more to the invention of the light bulb than just Thomas Edison. One thing is for sure that the inventor, born in Ohio, is still remembered as one of the greatest inventors of all time, and perhaps the greatest that America has ever produced.

❦

Although born in Ohio, Edison moved at an early age to Port Huron in Michigan. He was youngest of seven siblings and was born to a father who had to flee from Canada after he took part with the Rebellions in 1837. They eventually moved to Milan in Ohio where Edison was born on the 11th February 1847. He was seven when they moved to Michigan after the railroad business declined in Milan, and eventually, Edison would start working on the trains himself.

❧ II ❧

BORN INTO A COUNTRY
OF GREAT CHANGE

"There is no substitute for hard work."

— THOMAS A. EDISON

❧

Edison was born in 1847 into a country that was going through an incredible amount of change. The US is just coming out of a profound recession with railways taking off, and over the next few years, America would be connected coast to coast for the first time by rail. America was going through its industrial revolution including becoming one of the world's greatest producers of steel. The world was going through a rapid change as it had never been seen before, and Edison was born into a world that was looking for people like him who could progress society even further out from the dark and into the light.

Massive manufacturing and agricultural increases were happening at the time. The industrial revolution had already arrived in Britain, but this was America's turn for their own change, and it would also lead to substantial population increases. In 1840, the population of America was just 23 million, but by the end of that century, the population was more than tripled to 76 million. By the time of Edison's death in 1931, that figure would change to 124 million, which would eventually lead to today's figure of 325 million.

※

Edison was born into a country that was booming in all sorts of ways and finding its way from the independence that it achieved in 1776. When Edison was growing up, the states of America were really becoming united in more ways than the people who fought for their independence would have ever thought possible. It was changing from a country that had great potential, to the most powerful nation in the world that it is today.

※

Things changed quickly as they already had the blueprint from Europe of how to take a country into the modern age, except America had the advantage of learning from the mistakes so was able to achieve their revolution a lot more quickly. Edison was able to take the kind of ideas that had come out of Europe and improve on them in America.

EDISON'S DISABILITY

❦

Edison developed hearing problems as a young boy, which would affect him for the rest of his life. The exact reasons for the hearing loss are unknown, as Edison himself would enjoy making up tales about how it happened to him. In one story, he said it was because he got struck on the ears by a train conductor after his laboratory in a boxcar set on fire. Another story was that a train conductor had helped him get up onto a moving train by grabbing him by the ears.

❦

In truth, the reasons could be one of two things, and it is most commonly attributed to scarlet fever that he had as an infant, with ear infections that also went untreated. Another possible cause was that it was merely in his hereditary as he had a family history of hearing issues. Whatever the reason was, Edison didn't let it affect him, and he was still able to hear, but with great difficulty at times. When Edison was inventing the phonograph, at one time, he decided to actually bite down onto the instrument so that the vibrations would be delivered directly to his inner ear; he always found a way.

❧

Edison didn't let it get him down though, and even found reasons to be happy with his hearing loss as he said that it gave him a space in which he could think. He was offered the opportunity in later life to have an operation which could have improved his hearing, but he believed that it was going to affect his ability to concentrate in a world of much greater noise.

A DIFFERENT KIND OF
EDUCATION

❧

While many of the greatest minds in history went to some of the most prestigious institutions in the world, formal education isn't something that Edison neither needed nor wanted. From an early age, he was taken out of school and taught by his mother at home.

❧

In one of the greatest examples of being wrong of all time, Edison's teacher sent him home with a message to his mother that they thought he was "*addled*" and had a confused brain. Edison's mother immediately pulled him out of school after only three months of education. It was the making of Edison, and he would read everything that he could. He was devoted to his mother for her belief in him and desperately didn't want to let her down.

❧

The lack of formal education clearly didn't hold Edison back as his ever-inquisitive mind would give him all the answers to all the questions that he needed to know. He was the type of person to question everything, and if a teacher didn't know the answer to one of his questions, he's wanted to know why they didn't know the answer. It was that drive for answers that would see him be successful for the whole of his career.

⚜

At the age of 12, he would get his first job, working on a train, selling an assortment of various items to the people on board. He was never one for a classroom environment anyway, and with the help of his mother and his own drive for success, he was already on the right path to a successful career.

⚜

From his job on the train of selling such items as vegetables and newspapers, he'd go on to use the press on the train and create his own laboratory. This connected with his fascination of the telegraph as he knew the power that it could have. He developed a tactic of telling an operator how to telegraph headlines down the line, which would leave people wanting to read the news further down the line.

⚜

This would be a taste of thing to come with a man who not only would become an incredible inventor but also a brilliant advertiser and marketer as well. From an early age, he was already showing that he could drive up interest in something; it started with headlines, then he would use that ability to generate interest in all of his inventions.

8

❧ III ❧

THE TELEGRAPH AND THE START OF HIS GENIUS

"Being busy does not always mean real work. The object of all work is production or accomplishment and to either of these ends there must be forethought, system, planning, intelligence, and honest purpose, as well as perspiration. Seeming to do is not doing."

— THOMAS A. EDISON

❧❧❧

In the decades before Edison was born, electricity was being experimented with, but not many realized the uses it could have. Many people were fascinated with electricity at the time, and no more so that Edison who, at this age, decided to run a telegraph line from his home to of one of his friend's. It was an incredible feat for a man, but his incessant reading to further his knowledge gave him a great amount of skill.

∞

This was still the early days of telegraphy, and it was becoming more commonly used throughout the world. The first transatlantic cable would be laid in 1866 to link America to Britain and the rest of Europe. It was a time of change where communication was becoming more successful than ever, and it was something that Edison was keen to get involved in.

∞

A huge slice of luck happened for him out of near disaster when he saved the life of a telegraph operator's young child from the train tracks. It would put him on the right course to where he wanted to be as the operator showed his gratitude by training Edison to be an operator, which was a role that had a huge degree of status due to it being a very skilled position.

EDISON'S CURIOUS MIND

❦

A curious young man, Edison couldn't help but tinker with the instruments and make them more efficient. He wanted to increase the speed of the machines and make them better. He started to develop a modification to the apparatus, which not only improved the speed of the system but made Edison's life as an operator a lot easier too. He was still only a young teenager at this point.

❦

At the age of 15, he started to work for the Western Union who had a controlling interest in vast amounts of the telegraph system in North America. Between the ages of 16 and 20, he was employed as a traveling telegraphist by the Western Union until 1867. Edison got fired from Western Union as due to the experiments that he completed, he actually spilled some acid, and it dripped down to the desk below, which just happened to be of his boss. He had taken the night shift at the company so that he could work on such experiments, but spilling sulfuric acid onto your bosses desk is a sure way to get yourself fired.

☙❧

Shortly after this, he moves to one of their headquarters in Boston, which introduces him to an important set of resources including to some people who all think like he does and are keen to invent like he can. This is the start of Edison realizing that a collective of minds can achieve great things, and how he can use this collective to not only advance the technologies that he wants to but also being able to attach his name to them.

☙❧

During that time, there was a significant boom of invention, and it was almost a fashion to be an inventor. It could be said that one of Edison's inventions was the method of inventing. In Boston, he spends time in a shop near the Western Union office where a lot of inventors and entrepreneurs meet including Alexander Graham Bell who would be a great influence on him and also a later rival. At this time, he also helped to improve telecommunication with a series of small innovations.

HIS PATH TO STARDOM

❦

I t was at this time that he made a couple of breakthroughs that would see him on the way to becoming the inventor that we know today. One such breakthrough was with the stock ticker. It had been invented in 1867 by Edward Calahan, but Edison was able to make significant improvements to it. The improvements came with a printing telegraphy and a mechanism which allowed all the tickers to give the same information at the same time.

❦

His improvements also meant that an operator could bring all the tickers into line by sending them an electrical signal. This saved a lot of time and gained Edison a reputation in the business world as an inventor. The stock ticket he invented continued to be used for several years before it would eventually be replaced.

Around the same time, he was working on the quadruplex. A duplex machine was already in operation whereby two messages could be sent over one wire, but Edison had worked out a way to be able to carry four messages on the same wire and thereby doubling its efficiency. For all the minor improvements that Edison had made to the telegraph, this one was very significant, and it saved Western Union a lot of money as they could send more messages without building more lines.

❦

They loved it so much that they decided to buy it off Edison. It was the first big financial success of his career. He wasn't yet famous though, as you don't gain such fame by making improvements on stock tickers and telegraphs. It wasn't until Edison moved to Menlo Park that he would become a national celebrity.

❧ IV ❧

HOW EDISON CREATED THE INVENTION MACHINE

"There's a way to do it better - find it."

— THOMAS A. EDISON

❧

It was the 9,646th patent to be registered in the United States, and it was one to be able to count electoral votes accurately. It was the first ever patent to be registered by a Thomas Alva Edison, and it wouldn't be his last. 63 years later, the man would pass away, but not before registering another 1092 patents making him the most prolific inventor in history.

❧

His inventions were vast from the advancement of electrical light to the telegraph and the telephone. He invented the phonograph and was one of the fathers of the modern cinema. He moved from being in between various jobs to being a full-time inventor when he moved to New Jersey.

THE WIZARD OF MENLO PARK

❦

I
t was in 1876 when the man that we know today really
started to change the face of invention. He established
an industrial research lab, which was built in Menlo
Park, which was then a part of Raritan Township. But this
name has now been changed to Edison Township in his
honor. He built the facility out of the funds he made from
selling his quadruplex telegraph, and Edison was never shy of
investing his well-earned money into other projects.

❦

He sold the telegraph to Western Union for 10,000 dollars at
the time, which, in today's money, is nearly a quarter of a
million dollars. It was the first successful financial venture he
had and would pave the way for many more. Menlo Park was
thus created and became a factory of inventors. Many people
worked at the facility under the guidance of Edison, and
many of his great advancements came out of there.

One of Edison's genius moves was that no-one at the facility was able to register a patent under their own name. Everything that was done at Menlo Park was done under the name of Edison, which led him to claim credit for other people's work. While he was still heavily involved in all the inventions that came out of Menlo Park, it did mean that some of his associated didn't get any joint credit when they deserved it.

❦

Menlo Park eventually developed to the size of two city blocks, and it wasn't just a space for researchers to carry out experiments, it was also a vast warehouse, which would contain all of the chemicals and materials you'd ever need for the art of invention. That stock would include all types of animal hair and all other types of materials. It was an incredible place where minds were free to try and create incredible things, with everything right there at their fingertips.

❦

Menlo Park was the first research and development facility of its kind, and it's the base where Edison would have many of his most fabulous ideas. The first of his significant innovations at Menlo Park was the one that made him famous, the phonograph. By 1878, the invention was known all around the world, and it gave Edison his memorable nickname of the "***Wizard of Menlo Park,***" and the facility was known worldwide as the home of invention, with many visitors coming for a demonstration of the phonograph.

THE HOME OF THE LIGHT BULB

❧

Edison had his own name for Menlo Park as he dubbed it an *"invention factory,"* and it's clear to see why. Not only did it have the space available, but he also built an office and a library on the site, which was ever-expanding. Menlo Park was also the place where he progressed the invention of the light bulb. As 1879 turned over to 1880, on New Year's Eve Edison would give the road outside of Menlo Park, Christie Street, the honor of being the first ever street that was lit by incandescent light.

❧

It only took until the summer of 1880 for the bulb to be developed enough to be produced and sold in large quantities; he converted one of the buildings of the sight into a lamp factory so that he could keep up with demand. Menlo Park served as a testing ground for all kinds of electrical systems including having a network of underground cables that were able to light up lampposts around the site.

❦

It wasn't too long until Edison was lighting up homes as he established the Edison Electric Light Company and moved on from light bulbs to see what other uses he could find for electricity. He also created his own railway at Menlo Park, which ran on electricity; it ran all around the major states. In the meantime, the light bulb revolution would continue as he created a generator in Pearl Street in New York, which was able to light up an office and all the buildings.

❦

While at Menlo Park, he would apply for around 400 patents and people would come to the facility to see the work in action. There was a constant stream of businessmen and investors who wanted to see what the future looked like. While never stated as one of his inventions, the creation of Menlo Park and the platform that he gave to the invention is one of his greatest achievements.

❦

In 1881, he decided to leave Menlo Park and created a series of facilities across America. In 1887 though, he would build a new research laboratory that would become more of a permanent home for his mind in West Orange, New Jersey. This is where the leading research and development of his lighting companies would take place, and from this point, it is where he'd spend most of his time.

The research laboratory was massive, and Edison wouldn't have quite the same level of success or love for the place. It was too large for one man to oversee, and most of the innovations done at the facility weren't done under his own observation. It was at this stage where Edison became less of an inventor and more of a businessman.

V

EDISON AND THE PHONOGRAPH

"Just because something doesn't do what you planned it to do doesn't mean it's useless."

— THOMAS A. EDISON

The phonograph was the first of Edison's great inventions and the one that he was most proud of. It was able to record the spoken voice with the ability to play it back. Not to be confused with the record player, this was a separate device that would help pave the way for such machines. If you didn't know, it'd take many guesses to work out what Edison first said into the recording device, which was his own rendition of "***Mary had a little lamb.***"

The phonograph isn't remembered very much today, but it is perhaps the most excellent work of Edison. It was invented in 1877 and was the first ever device that could reproduce recorded sound. It's quite difficult to imagine how big of a news story that would have been at the time, imagine hearing recorded sound for the first time ever!

❧❧❧

It was a huge deal, and Edison toured the device around the country, and it created a buzz wherever it went. He actually took it to the White House, and there is a story that the President of the time Rutherford B. Hayes kept Edison company until 3 am and woke up his wife, this is how much he was fascinated with the device.

❧❧❧

It worked by having two needles, one to record the sound and another to playback that sound. The recording needle would recognize the vibration of the sound coming through the mouthpiece, and then the vibrations of that sound would be recorded on a cylinder. In the early days, the cylinder was made out of brass before being wrapped in tinfoil.

❧❧❧

The invention wouldn't be developed any further for a long time as Edison became a lot more focused on the incandescent light bulb, which he'd become known for. The existing phonographs couldn't really be sold as the tin foil was too fragile to be of any use. Finally, after 10 years, Edison turned his attention back towards the phonograph.

※

Edison improved the device by using wax cylinders instead, which recorded a better sound and were also a lot more reliable. The devices were then available for mass sale and were sold for a price of $20, which was still very expensive in the 1890's, and they could only record music for about two minutes. Eventually, the industry moved on from Edison's invention as the use of discs rather than cylinder came a lot more popular.

※

As a legacy though, Edison had produced the first recorded sound and made the first records company. It's a legacy that he is not very well known for today, but as you put in your earphones, hear the television or the radio just remember that it was Edison who first ever listened to a recorded sound, and he shared that gift with the rest of the world.

OTHER NOTABLE INVENTIONS

❦

Motion Picture – Often seen as the father of the motion picture, it was one of Edison's most celebrated works. The first device he created looked much like his phonograph machine. As you can imagine, it wasn't a very usable or workable machine, but it did allow him to develop the next stage. His next device worked much in the same way as those classic film reels of years gone by. A series of pictures were fed through a machine to create a movie. At the first stage, only one person at one time could see the movie, but this changed. It's another example of how Thomas Edison carved the world around us.

❦

Vote Recorder – This was to be the first of his many patents, and he was just 22 years old when he filed for it. It aimed to help legislators who were in Congress to record their votes more quickly than the usual method of the time, which was by a voice vote system. The names of the legislators were embedded into the system, and they would move a switch to either record a yes or a no vote which would send an electric current to a clerk's desk where the votes were counted.

⚜

Electric Stencil Pen – If you have a tattoo on your body, then it's a good chance that one of Edison's creation inspired the device that made it. The pen was made with the intention of perforating paper which was helpful in the printing industry and could help copy documents more efficiently than ever. That idea was taken onboard by Samuel O'Reilly who created the tattoo machine, based on Edison's invention.

⚜

Electric Power Meter – If you have a meter in your home that records your electricity use, you have Edison to thank for that as well. The problem apparently arose after electric cables were fed through into each home. There was no way of recording how much each house used, and therefore, no way of knowing how to bill them. He created a solution whereby zinc would travel from one transmitter to another at a set rate. The meter reader would be able to determine how much was used and how much the customer should pay.

⚜

Fruit Preservation – After perfecting the method of making a light bulb into a vacuum, Edison took it to open himself to see what other uses he could have for an airtight container. One of the ways way to help with the preservation of fruit whereby it would be placed in a glass container before all of the air was sucked out of it. While it may have worked, it was more trouble than it was worth for, so people didn't really catch on. It's another look into the mind of a man who was always thinking about how to make the world easier.

⚜

Car Batteries – Edison was way ahead of his time when it came to thinking about cars. They weren't even being mass-produced, and Edison was thinking about putting electricity into them. He decided to make an alkaline storage battery with the intention of running a car for at least 100 miles. The technology wouldn't match up with his mind though, and cars continued to run on gas. I'm sure he'd be proud of the battery technology in cars today but perhaps annoyed that the most well-known all-electric car company in the world was named after his fierce rival, Tesla.

⚜

Concrete house – It doesn't sound too comfortable, and perhaps that's why it didn't work. The plan behind it, however, was a noble and gracious one for Thomas Edison. He wanted to create a house that could be affordable for those on low incomes. He planned the homes to be around one-third of the average price at the time. He couldn't get enough investors to buy into the project, and it never really got off the ground. It was too much of a radical step for anyone to take.

⚜

Concrete furniture – Inside his concrete house, Edison wanted to put furniture made of concrete. Again, the aim was a noble one for him to make cheap and affordable furniture for those who couldn't afford it. He wanted to provide them with a long-lasting substitute that wasn't going to break. The reason it failed though was clear. Having concrete furniture would be heavy, uncomfortable, and ugly. It was a practical solution to a problem, but it was one that people were never going to be interested in.

***

Phonograph dolls – With Edison a lot of the time you get the impression he was a man who was so excited about what the future could bring that we wanted too desperately to bring it into the present. That can be seen with the phonograph doll, which would play back nursery rhymes to little children. The problem though was that the technology was unreliable and the sound quality was terrible. It paved the way for the thousands of talking toys that we have today.

***

Magnetic Ore Separator – This was built with the intention of separating low-grade ore. This would allow previously discarded mines to be reopened as Edison could extract iron ore from rock more efficiently than ever before. He invested a tremendous amount of time and effort into the project that could well have worked, but the price of iron ore fell, and Edison had to abandon a project that he thought was going to achieve great success.

❧ VI ❧
IT WASN'T ALWAYS ABOUT SUCCESS FOR EDISON

"The three great essentials to achieve anything worthwhile are: Hard work, Stick-to-itiveness, and Common sense."

— THOMAS A. EDISON

❧

Edison wasn't afraid to take risks, and he also wasn't scared of failure. He knew that with inventing came the chance of things not working the way that he wanted them too, some of his ideas were ahead of their time, and some were just not able to sell. He once said

"I have not failed 10,000 times—I've successfully found 10,000 ways that will not work,"

which shows his mindset when it comes to developing a new technology.

❀

One of his first failures was with the electronic vote recorder, which he invented while still working for Western Union in 1868. The invention meant that officials would be able to cast their vote on the machine that tallied them automatically. He knew that the device would save many hours for the public officials that would use it and also thought it was going to be the first inventions that earned him a nice sum of money.

❀

Instead, they didn't want to know. The world of politics was, and still is, a murky one and legislators didn't want a device that could affect the vote trading and manipulation that could be done under the current system. It was an early lesson for Edison, not with politics, but also invention as it was a lesson he applied for the rest of his life. He was no longer going to invent something for the sake of invention. Instead, he was going to invent things he could sell. It was a lesson that would gain him spectacular wealth, but he still wouldn't be without failure.

❀

The electric pen was such an invention that he thought would have its uses as the railroad companies had a market for tools that could speed up any process. Edison wanted to make it easier for handwritten documents to be copied, so he invented the pen that had a small electric motor and a battery. The pen wouldn't deliver ink but instead would punch small holes through the surface of the paper, which could create a stencil on wax paper that then could be covered in ink and pressed on blank pieces of paper.

⚜

Then pen wasn't a disaster though, but it was a bit too ahead of its time as the noise and weight of the pen made it unpopular. The batteries had to be maintained by using a chemical solution, so it was a very messy affair. As other inventions were taking off, Edison decided to abandon the development of the pen and move on. Perhaps the most significant legacy that it has is the inspiration behind the tattoo machine that proceeded it.

⚜

In 1887, Edison had opened up a new laboratory in New Jersey and wanted to raise funds that he could put back into the facility. His idea to raise those funds was to have a change of tactics from his usual line of thought and develop products that didn't require much thought, but ones that he thought would turn into a good profit.

⚜

One of these inventions was the talking doll. He imported the dolls from Germany and made a smaller version of his phonograph. His idea of 'get rich quick scheme' didn't work, and the toys quickly developed a series of issues. The toys were too fragile, the voice of the doll didn't last, the voice would also sound unpleasant, and Edison quickly withdrew them from the market. Just like with the electric pen, it was perhaps the inspiration to others, which is the greatest legacy for his phonograph doll.

THE BIGGEST FAILURE OF ALL

❧

While we have only talked about small and non-damaging inventions so far, if Edison could go back and have his time again then surely he would never try and get into the iron ore industry. For ore to be smelted, the surrounding nonferrous rock had to be removed. Edison developed a system which could achieve this on an industrial scale and wanted to process 5,000 tons a day.

❧

The system though encountered a series of expensive problems. He spent countless hours trying to redesign the crushers, elevators, and other types of machines. While his other failures weren't too bad and didn't take up a great deal of time, this one was awful. It didn't damage his reputation too much, but it did ruin his pockets as it was incredibly expensive, and he didn't let go of the project until 10 years after starting it.

Edison also wanted to experiment with X-rays and see what innovations he would make with the technology and left the research to one of his associates, Clarence Dally, who completed multiple experiments using this new technology. Neither Edison nor Dally had any idea of the risks involved, which would lead Dally to have severe burns on his arms and sores all over his body, he would continue to get increasingly ill, and Edison kept him on the payroll, even when he could no longer work. Dally eventually passed away due to his radiation poisoning, and Edison never went near x-rays ever again.

❧

It's perhaps his failures that show how Edison achieved what he did in his life. He wasn't afraid to fail, and that lack of fear led to some of the greatest inventions and advancements of all time. If he didn't try and invent the electric pen, the tattoo machine wouldn't have been invented, and if he hadn't built a phonograph into a doll, then it wouldn't have inspired the millions of toys that would follow.

❧

If he was scared of failure too, then he wouldn't have had the confidence to achieve all the great success that he did. For all his failures, his successes overwhelm them. When you take as many shots as Edison did, then some of them are bound to miss. In the end, he will forever be remembered for what he did do rather than what he didn't.

❧ VII ❧
THE TRUTH ABOUT THE LIGHT BULB

"Discontent is the first necessity of progress."

— THOMAS A. EDISON

❧

Despite what a lot of people believe, Thomas Edison did not invent the light bulb. There are a quite a few inventions that aren't very clear regarding who actually invented them, same goes for the light bulb, which has gone through numerous changes and improvements until we get the product that we all use today.

❧

What Edison did was create the first incandescent light that could be made available to the public on a commercial scale.

Numerous people had created their own version of the light bulb before Edison came along, but no-one was able to work out how to create a light bulb that everyone could use.

※

The reason that Edison is often mistakenly called the inventor of the light bulb these days is the same reason that many people think Henry Ford invented the car. What both men did was create a way for a product that was already invented so that it could be sold to the masses. There was no point in the light bulb existing if people weren't going to be able to use it, Edison found a way.

THE REAL INVENTORS

꧁꧂

There was a man called Humphry Davy who invented the first electric light in 1802, and it wouldn't be for another 78 years that Edison's light bulb would go into production. That invention by Davy, however, was by accident and it wasn't practical. Having connected wires to an electric battery with a piece of carbon, the carbon glowed producing light. That light bulb would come to be called an Electric Arc Lamp.

꧁꧂

In 1850, the first 'bulb' was created by Joseph Swan, which was done by placing carbonized paper filaments in an airtight glass bulb. There were problems though, and the light wouldn't last for very long at all. He continually tried to develop his light bulb and made some great strides, but still wasn't able to fully crack the code.

It wouldn't be until 1878 before Edison seriously got involved with the light bulb and filed his first patent of it on the 14th October called "***Improvement In Electric Lights,***" which pushed him on the right path of achieving his dream of a commercially available electric light bulb. As with the inventors before him, Edison attempted countless amount of materials as the filaments in the light bulb,o and he and his team worked tirelessly to find it.

BAMBOO WAS THE ANSWER

✦

Eventually, the discovery was found, and it came from an unlikely source. The material he used was a stand of carbonized bamboo, which could shine for longer than any other material. This was the beginning of light bulbs being in every home, and in 1880, his Edison Electric Light Company began marketing a product that would change the world forever.

✦

While the idea of having an electric light encased in a bulb wasn't Edison's, and that's not to distract away from the importance of what he did. In many ways, you could say that Edison invented the light bulb as he invented a way for it to be usable. He took the basic premise and made considerable improvements and worked tirelessly until a solution was found.

He and his team eventually found a solution that would be one of the most groundbreaking discoveries that have been ever made. The basic bulb that works in most homes today still comes from Edison's initial design and has only ever been improved on recently with the LED light, which can last for much longer than traditional lights.

❧ VIII ❧
THE RIVALRY BETWEEN TESLA AND THE "WAR OF THE CURRENTS"

"The chief function of the body is to carry the brain around."

— THOMAS A. EDISON

❧❧❧

Thomas Edison and Nikola Tesla, both men were geniuses, but there was a rivalry there, which helped drive each other to achieve greatness. Today, when we think about competition, we think about sport and the big showdowns that can happen. In the 1880's, though two men would go head-to-head in the battle of AC/DC. For anyone who doesn't know, the famous Australian rock band's name comes from two types of current, alternating current and direct current.

❧❧❧

Edison was very much a proponent of direct current and Tesla was in the other corner supporting his favorite alternating current. It was a *"war of currents,"* but who would come out on top? It was about the idea of which type of current would be best for the new electrical era that was well on its way.

***

The two were very different in their methods and their ideas. Edison was a practical man who would achieve his success through trial and error with careful planning. Making step-by-step improvement until a solution was found. Tesla was more of a maverick who would have wild dreams and tried to realize them. They were to minds of genius that worked in two very different ways.

***

Edison's direct current was able to achieve a lower voltage from the power station to the consumer, so Edison declared that it was much safer than any other method. Tesla, on the other hand, argued that his alternating current would be able to travel over much larger distances as the flow of energy could change direction and alternate.

***

Tesla pleaded with Edison to give him a chance to prove it, and Edison bets him $50,000 that it could not be done. When Tesla tried to claim his bet, Edison declared that it was a joke saying

"When you become a full-fledged American, you will appreciate an American joke."

The Serbian-born inventor wasn't amused and quit Edison's company; he eventually saved up enough to form his own Tesla Electric Light Company, which obviously used AC current.

❧

At this time, a man called George Westinghouse, who had previously invented the railroad air brake, decided to create a company that was going to compete with Edison. Westinghouse recognized the genius of Tesla and brought him on board and bought that patents that Edison has previously dismissed. By 1903, Tesla was harnessing the power of the Niagara Falls and transmitting that power all the way to New York, which showcased the ability of AC current to travel long distances.

EDISON'S GRUESOME
EXPERIMENTS

❦

One of Edison's misadventures was to try and discredit the AC system with all his power. He had invested a lot in DC, and a change in current would change the way that he fundamentally operated, and this was something that Edison was prepared to fight for and ended up doing so in some horrible and gruesome ways. One of those ways was by developing the electric chair after New York State wanted a more humane form of execution than hanging.

❦

They commissioned the world's first electric chair, which was to be powered by three generators from Westinghouse, obviously using AC as the current. The chair gained a lot of negative press for Westinghouse, and he tried to block their use via a court order. In 1890 though, a man by the name of William Kemmier had the so-called humane honor of becoming the first ever man to be executed via the electric chair.

⚜

It was a gruesome death, and the electric chair was never able to provide the humane death that it was originally meant for. That was the first of over 4,000 deaths using the method. Thankfully, the electric chair is barely used today as none of the states has it as their primary method of execution. The demonstration was effective in showing how dangerous AC was, but DC would have been equally lethal anyway.

⚜

In another show, Edison also got the local children to collect stray animals that he could use for his experiments, hooking the animals, mainly dogs, to different types of electrical current. He would electrocute the dogs with DC, and they would still be alive, and then would kill them with AC. It was a horrible way to prove a point and shows Edison at his most driven and most ruthless.

⚜

Edison continued his campaign against AC by other cruel means. Topsy was a mistreated elephant who had a series of incidents where the animal had killed or injured humans. The park where she was kept decided they could no longer look after the animal and announced that it would be hung, however impractical that seems. Instead, Edison thought it would be a good idea to electrocute the elephant, again to show how dangerous AC was.

Edison decided to film it into a short film, which was to be distributed by his Edison Manufacturing Company. It was 74 seconds long and showed the execution taking place. It required a lot of clever manipulation as not only was the elephant electrocuted, but it was also fed carrots laced with cyanide and strangled once it fell to the ground. By this time, the 'war of currents' was effectively already lost for Edison, and he eventually had to concede.

❧

AC was simply not as dangerous as he said it was, and it went on to be the more commercially used current. Edison, though, was right in many ways. DC is safer and more practical when it is used over short distances. You know the laptop cord that comes in two with the little back box in the middle? Well, that is there to convert Tesla's AC current back into Edison's much loved DC current. In many ways, both men were right. AC is the current that runs into your home, but all your electronic devices convert that current in DC.

✣ IX ✣
FATHER OF THE MOTION PICTURES

"Non-violence leads to the highest ethics, which is the goal of all evolution. Until we stop harming all other living beings, we are still savages."

— THOMAS A. EDISON

❦

In 1888, Edison's laboratory developed a device that would come to be called kinetograph. Edison wanted to do to the eyes what he had done to the ear with the phonograph. It paved the way from the basic principle for the moves what we watch today whereby a series of quick images are taken and then played together to make it look like a moving picture.

❦

He and his team designed a long flexible strip of film that would be fed through a machine and played back. They built a device called a kinetoscope, which could play these images, and people would pay a small fee to watch them. However, at the start, only one person would watch at any one time.

❦

From this, Edison created his own production company and would take short films to be sold on. Edison bought other inventions such as projectors, which helped advance the movies that his production company could show. He created America's first movie studio in 1893 called the Black Maria where he would film performers on a stage and sell the videos.

❦

Not only did Edison help advance the movie industry to what it is today but also has a curious claim of possibly being the first movie pirate of all time. He would send his associates out to Europe to collect movies that were being made and then sell them in America as if they were one of his own.

❦

While Edison didn't invent the first motion picture, what he did bring is the public imagination in America. Just like with the light bulb, he had taken on something that existed, made it a lot better and found a way of earning money from it. A lot about the film industry we know today could be traced back to Edison.

❦ X ❧

THE MAN BEHIND THE INVENTIONS

"The reason a lot of people do not recognize opportunity is because it usually goes around wearing overalls looking like hard work."

— THOMAS A. EDISON

❦

It was a Christmas day that Edison would marry the love of his life in 1871 when he married a 16-year-old woman by the name of Mary Stillwell. They had only known each other for two months, and 24-year-old Edison fell in love, which an employee at one of his shops.

❦

They went on to have three children together, with the first birth being when Mary was still only 18 years old. The three children were Marion Estelle, Thomas Alva Jr., and William Leslie who would all go on to live long lives, with the youngest William being an inventor himself, graduating from Yale in 1900.

※

Unfortunately, however, his wife Mary would not go on to live a long life and would die tragically at just 29 of unknown causes. It is thought that the cause of death could have either been from a brain tumor or a morphine overdose. At that time, morphine was prescribed by doctors for some various causes, and it could have been too much for her body to take.

※

Edison was a driven and obsessive man who would more likely be seen in the laboratory than with his family. His families' loss was the worlds gain considering the impact that he had on the world. Edison would marry again, however, when he was 39 years old, again to a much younger woman. This time it was to the 20-year-old Mina Miller who would go on to eventually outlive Edison as she passed away in 1947.

※

When they first started dating, Edison actually taught Mina Morse code. Despite Edison's long hours of work and his dedication to his profession, their relationship was a very loving one, and when Edison came to propose to Mina, he did so by tapping out

"will you marry me?"

on the palm of her hand. Mina had thankfully remembered what Edison had taught her and understood the message, responding back with a *"yes."* They also had nicknames for each other with Mina calling him *"dearie"* and Edison calling her *"billie."*

❧

One story does show the balance that Edison had between love and invention when he combined the two to do something for his wife. Mina was a lover of nature and especially bird watching. In the winter, however, the water in the bird feeders outside their bedroom would freeze over, and therefore, the birds would not come. Edison decided he would solve this issue by feeding an electrical line into the birdhouse so that the water could be heated. There was a switch in the bedroom that could be flicked so that Mina could continue one of her favorite hobbies.

❧

Edison had three more children with her, and that's where it would end for Edison, with his 6 children in total, with Madeleine, Charles, and Theodore Miller adding to the three that he had with Mary. All three of his children with Mina had notable lives. Madeleine would go to marry one of the pioneers of airplane manufacturing in John Sloane, Charles became the Governor of New Jersey from 1941 until 1944 and would take over his father's company after his death, and Theodore would graduate from MIT in physics and would go on to register over 80 patents himself.

HE LIVED FOR HIS WORK

❧

E dison was a free spirit in many ways and a free spirit who seemingly only wanted to do what he wanted to do. This would involve working for over 90 hours a week at times, and he'd expect the people who worked for his company to be as equally devoted to their work. He enjoyed his role as an inventor and also enjoyed having a team around him who were equally of the same mindset who would help him achieve his goals.

❧

He wasn't one for meaningless social engagements, even when his wife Mina would arrange formal dinners that the couple would host at their Glenmont mansion in New York. There were often times where he would feign indigestion to get out of such engagements; sometimes he pretended this indigestion even before he had eaten any food. He would make his way through the kitchen out of sight, grab some food to feed his appetite and then retire to his living quarters by sneaking off up the servant's stairs so that he could carry on inventing.

❦

He did sometimes find the time, however, to mix his work with his family life. He would often use his children to get reference materials from their library. The children would go off, find the information he needed and create a note on paper so that he could read the information that he needed. He had desks at his home with Mina that were side-to-side whereby he could study, and she could deal with all the social elements of Edison's life.

❦

At the mansion, approximately 4-6 servants were working there at any one time. If they were hosting a big event, then there was room enough for that number to quadruple. The working conditions at the Edison mansion weren't severe at all as the pay was good and there were also room and board available. Edison affectionately referred to his staff as the "league of nations" on account of the number of different nationalities between them.

❦

At one time, a number of the staff were from Scandinavia and on one Christmas Mina ordered in a Swedish candle box Christmas tree to honor their hard work at the mansion. The Edison family also had no special requirements on the staff sleeping on different floors, and their son Charles actually occupied what was meant to be a guest bedroom.

EDISON KEPT GREAT COMPANY

᭤᭟᭤

The mansion was a popular residence and hosted a number of well-known guests including most frequently Henry Ford and his wife, Clara. Other notable guests were the Kings of both Siam and Sweden as well as another great inventor in Orville Wright. Also at the property stayed presidents Hoover and Wilson, which shows what kind of regard Edison was held in. While not always being the most sociable of figures, Edison did love surrounding himself with great minds of the time.

᭤᭟᭤

Edison tried to keep his religious beliefs quiet and would rarely discuss them. He once said

> *"Nature is what we know. We do not know the gods*
> *of religions. And nature is not kind, or merciful,*
> *or loving. If God made me, the fabled God of the*

three qualities of which I spoke: mercy, kindness,
love; he also made the fish I catch and eat. And
where do his mercy, kindness, and love for that
fish come in? No; nature made us, nature did it
all, not the gods of the religions."

which makes it sound as though Edison was an atheist, but he later distanced himself from that idea.

❦

Later he would say that

"You have misunderstood the whole article because
you jumped to the conclusion that it denies the
existence of God. There is no such denial, what
you call God, I call nature,"

and while he didn't like to talk about it, it does seem as though Edison believed in a supernatural being, but not God as such.

❦

Despite his showing with the elephant and his advancement of the electric chair, Edison wasn't a man who believed in violence in general. He was once asked to serve as a Naval Consultant during World War 1 but would only do so if he was able to work on defensive weapons later stating that he was proud to have never invented weapons to kill.

❧ XI ❧
EDISON'S GREATEST QUOTES

"The best thinking has been done in solitude. The worst has been done in turmoil."

— THOMAS A. EDISON

❦

Edison was a man who was able to deliver some of the greatest quotes of all time that can still be applied today. Some of them are funny, while others show his dedication to hard work. He would often use the sound bite to promote his inventions and himself.

❦

"Our greatest weakness lies in giving up. The most

*certain way to succeed is always to try just one
more time."*

This quote is an insight into the mindset of the man, his drive
to find a filament for the light bulb was relentless and tried
hundreds of different ideas.

❧

"I start where the last man left off."

Again, while Edison is famously credited with the invention
of the light bulb, he simply picked up from where the last
person left off and made something better. A lot of Edison's
greatest achievements were advancements on other work.

❧

*"Just because something doesn't do what you planned
it to do doesn't mean it's useless."*

The mind of an inventor is to think about things from new
angles that no one else could have thought of. Edison was
able to learn from his mistakes and used those lessons to
make his inventions even better.

❧

*"If we did all the things we are capable of, we would
literally astound ourselves."*

Edison was a man who had complete faith in himself if he
worked hard enough. His belief most likely came from his
mother who believed in her son, even when his school didn't.

"When you have exhausted all possibilities, remember this: you haven't."

Who would have thought to use bamboo? Edison showed with his discovery that he was willing to go to lengths that no one else would go. When others would lose hope, Edison wouldn't.

"Opportunity is missed by most people because it is dressed in overalls and looks like work."

A lot of Edison's quotes are around the premise of hard work. He wasn't a maverick genius, but a genius of the method - trial and error. It wasn't glamorous at times, but it worked.

"The three great essentials to achieve anything worthwhile are hard work, stick-to-itiveness, and common sense."

They aren't groundbreaking essentials, but Edison had each of them in abundance. He would never stop until he got the answer he wanted, sometimes to his detriment.

"Genius is one percent inspiration and ninety-nine percent perspiration."

Perhaps his most famous quote. All the memorable geniuses we know today worked extremely hard. While their minds may have worked in ways that most people's can't, they couldn't have achieved what they did without hard work.

🕮

*"Results! Why, man, I have gotten a lot of results. I
know several thousand things that won't work."*

According to one of Edison's associates, around 1,600 attempts were made for a light bulb filament before the correct one was found. You could count that as 1,599 mistakes, but they all didn't matter when that one success came.

🕮

*"Many of life's failures are people who did not realize
how close they were to success when they
gave up."*

To achieve anything, you need perseverance, and no one knew that more than Thomas Edison. We would keep going until he found the answer.

🕮

*"Everything comes to him who hustles while
he waits."*

Perhaps one of his more philosophical quotes, Edison's mind never really switched off the task, and he was always thinking about his work.

"Being busy does not always mean real work. The object of all work is production or accomplishment, and to either of these ends, there must be forethought, system, planning, intelligence, and honest purpose, as well as perspiration. Seeming to do is not doing."

Edison's invention factory at Menlo Park was like a production line of ideas; it enabled everyone a great space to think in which to be productive at all times.

"Your worth consists in what you are and not in what you have."

This is a wise quote from a man who did have everything. He didn't let his money change him though and continued to be an incredibly hard worker.

"I never did anything by accident, nor did any of my inventions come by accident; they came by work."

As with any inventor, there are times where Edison had his slices of luck. You only get to that position through hard work, and he deserved all of the luck he got.

"I have not failed. I've just found 10,000 ways that won't work."

It's hard to describe something as a failure if it is just a means to success. Edison used the trial and error method on numerous occasions to achieve what he wanted to.

❧

"To have a great idea, have a lot of them."

Imagine all the terrible ideas Edison had. His mind would have been working overtime at all time, and I'm sure he had a lot of thought for ideas that were terrible, then ever so often, a great one would come into his head.

❧

"One might think that the money value of an invention constitutes its reward to the man who loves his work. But...I continue to find my greatest pleasure, and so my reward, in the work that precedes what the world calls success."

Edison never stopped, he made a fortune at an early age, yet never lost that desire of will for invention, and he never let his money change his attitude.

❧

"There's a way to do it better. Find it."

This quote makes you wonder what he would think about the world we know today. A world of LED light bulbs, smartphones and 3D cinema. He found a better way to do almost everything, and the world took on that baton and improved it even more.

❧

"What you are will show in what you do."

What did Edison did was leave a legacy of two woman and six children who loved him dearly. He also left behind a world that was much improved from the one he found.

❧

"I never did a day's work in my life. It was all fun."

A great quote from a man who worked incredibly hard, but did so with a purpose and an aim. His drive made him become one of the greatest inventors of all-time.

❧

These quotes show a man who had all his priorities in place. He knew the value of hard work, but also knew that there was a lot more to life than electricity. He continues to be an inspiration to many, he earned all of his opportunities and then made the most of them.

THE TIMELINE OF EDISON'S LIFE

❧❧❧

- 1847 – Edison was born on February 11 in Ohio.
- 1859 – At the age of just 12, he starts selling newspapers and other items on a train that went through Port Huron, Michigan, and Detriot. He would continue to do this for four years.
- 1864 – Starts working as a traveling telegrapher before ending up in Boston.
- 1868 – Registers his first patent, which was an automatic vote counter.
- 1869 – Works on a stock ticker, which would be his first source of significant income.
- 1871 – Marries his first wife, Mary Stillwell, on Christmas Day. They would have children in 1873, 1876, and 1878 before she later died in 1884.
- 1874 – Invents the quadruplex telegraph, which he would sell to Western Union later that year for $10,000.

- 1876 – Opens up Menlo Park in New Jersey that would be the base for many of his great inventions.
- 1877 – Invented the first basic phonograph, the invention that first made him famous.
- 1879 – Creates the world's first useable incandescent light and would continue perfecting the idea, later famously generating power from Pearl Street Station in New York in 1882, the first commercial central power plant in America.
- 1885 – Nikola Tesla quits Edison's company, effectively starting the ball rolling what would become the "war of the currents" two years later.
- 1886 – Marries his second wife, Mina Miller, following the death of Mary two years earlier. They would have children in 1888, 1890, and 1898. They were married until Edison passed away.
- 1887 – Opens up his new research laboratory in West Orange, New Jersey.
- 1888 – Develops the first motion picture.
- 1892 – Creates the company Edison General Electric.

❧ XII ❦

HIS DEATH AND THE LEGACY HE LEFT BEHIND

"Show me a thoroughly satisfied man and I will show you a failure."

— THOMAS A. EDISON

❧❦

In his final years, Edison's health began to suffer as he struggled with diabetes. He would end up dying due to complications with the condition on October 18, 1931. He died at his home in New Jersey and would later be buried just behind it as well. He was 84 at the time and was still working for as long as he possibly could.

❧❦

Quite curiously, there is a test tube at The Henry Ford

Museum near Detriot, which apparently contains the last breath of Thomas Edison as apparently Ford convinced Edison's son Charles to seal it in a tube as a monument to his life. A plaster death mask was also made of his face as well as casts of his hands.

৩১৫৩

In a touching moment after his death, many people and companies around the world dimmed their lights for a minute in honor of the great man who made it possible for light bulbs to be shining all around the world.

THE LEGACY OF EDISON

※

E dison's legacy is one of a brilliant man who changed the world and made much advancement to bring forward the world that we see today. He is credited with things that he didn't achieve, but what he did accomplish was incredible. He will forever be most associated with the light bulb, and for a good reason. Most advancements are pivotal to the world that follows them, and Edison's devotion to perfecting the incandescent light bulb enabled society to move on and create an even more incredible world.

※

While many of the greatest minds of all time were reclusive or at least introverted, Edison was a man who understood the value of what everyone else around him could help him achieve. With that team, he was able to do incredible things. Mass production has only recently been delivered to the industrial world, and Edison took these same principles and applied them to inventing, allowing great minds to come together to achieve great things.

❧

This was a rags-to-riches story and one of the greatest to be told. He was the personification of the American dream; his father had to flee out of Canada to Ohio, he developed hearing problems, his teacher said that he was addled, and he started working before he was even a teenager. To overcome what he did and to achieve what he did was an incredible achievement.

❧

There are times when he wouldn't be a nice man, he demanded a lot from his employees and would happily crush his rivals like he tried to do with Westinghouse. He had an enormous ego, but one that was justified in many respects. He had a brilliant mind and a large family around him.

❧

The way that the man is often portrayed in the modern day is one of the most complicated man who doesn't deserve the credit that he gets. His popularity has decreased, and at times, it seems like it has almost become a fashion to discredit him and point to all the negatives aspects of both his character and his inventions. The same people seem to love talking up Tesla and what he achieved.

❧

It is true that Edison is wrongly attributed to some inventions, and it is true that he took credit at times for work that wasn't his, but it'd be wrong to stop there. What needs to be remembered is what the man actually did achieve. The advancement of the telegraph, the invention of the phonograph, and taking light bulbs to the masses to name just a small few.

⁂

Despite the negative press at the time, Edison's legacy is secured for a long time to come. If it weren't for him, we just wouldn't be living in the same world. Even if you consider what he took credit for at Menlo Park, nothing would have been made if he hasn't built the facility and bought all those minds together.

⁂

Edison helped accelerate the world into a new era. As a man, sometimes his legacy is one of a difficult man who had an absolute drive for his work. As an inventor, he should rightly be remembered as one of the best of all time.

THE STRENGTHS AND
WEAKNESSES OF THOMAS
EDISON

৩১৩

Edison always had the values of hard work within him and would do so until the day that he died. This can already be seen from an early age after his mother had pulled him out of school, Edison was keen to read and take on all the knowledge he could. While other children would have had their distraction, Edison would have had his nose into a book.

৩১৩

This dedication to hard work led him to be working on the trains at 12 years old and would eventually get him a job for Western Union. Not content with the working life, Edison experimented while on the train and continuously looked for ways to improve the world around him. Once he earned a handsome sum of money, he didn't stop and used the funds instead to build Menlo Park.

From there, he would create a factory of invention and take his work across the country, before building another research facility. He would often work in 60 hour spells only taking a cat nap in between and would often work from the moment he got up till the moment he went to sleep. That hard work helped him to start from nothing to becoming a man still remembered well over 80 years after his death.

His story is also one of overcoming the difficulty. From a very early age, he had developed hearing loss, and many could have seen it as a sign to give into a challenging world where he struggled with one of his senses. Instead, he saw it as a blessing and enjoyed the solitude that it provided. Edison never let his disability hold him back, which is even more incredible considering he was the man who invented a device that was the first ever to record sound.

Edison also shows us the value of teamwork. What he created at Menlo Park was a stunning example of great minds getting together to achieve something special. These days, it seems so commonplace with great breeding grounds of thought at the likes of Facebook, Apple, and Google, but in Edison's day, it was the first of its kind. They were able to create products at an incredible speed and allowed America and the rest of the world to speed ahead into the modern world.

Edison's genius did come at a price though - with his obsession with certain things that he just couldn't let go. Idea's like his magnetic ore separator took far too much of his time, and his rivalry with Tesla and the 'war of the currents' were an example of a time where he didn't know he was beaten. That obsession did also lead to some great inventions though.

⚜

Edison loved his work, but whether his family loved it was another matter. He would often spend long periods away from home and the hours that he worked meant not a lot of quality time for the family. His work-life balance tipped almost entirely in favor of work.

⚜

It may have been a strength in business, but Edison was a ruthless man in many ways and expected a lot from the people around him. He took credit for all the work completed at his facilities, and his ego was an incredibly big one, he was a prominent character who engineered everything about his life to be all about him. Many people weren't credited with incredible work, but Edison was able to walk away with the praise that he didn't deserve.

HOW CAN WE USE EDISON'S
STRENGTHS IN OUR LIVES?

As a rags-to-riches story, there's not a lot better than Edison's. For all his faults, the man shows the value of hard work in every possible way. While people may talk about his being falsely remembered for inventing the light bulb or taking credit for all the invention at Menlo Park, it's easy to forget what the man achieved. He was a great mind, but none of that would have got anywhere if it wasn't for his drive and determination. Imagine - what if Edison decided not to work hard? It would have been a great mind gone unused. It makes you wonder what you are capable of if you had the same dedication.

⚜

It's not just the dedication though; it was the ability to use that hard work to think big and never settle for one thing. He pushed himself to achieve more and wasn't afraid to take risks. He was sacked from Western Union and went to create his early inventions and wouldn't let his disability hold him back.

While not many could ever dream of the man's ego, he did show what could be achieved with a little confidence. He made bold claims and did everything he could to the back it up. Edison showed that if you can just believe in yourself, then you will be able to achieve things that others would have previously have thought impossible.

He also taught the world the value of teamwork. He worked very well in a group and creating an environment where everyone could feed off each other was a brilliant way to get things done at an amazing pace. Great minds shouldn't always be left alone to think; sometimes they need a soundboard and someone to bounce off. Edison was never afraid to take on someone else's ideas and use them for the sake of invention.

❧ XIII ☙

REMEMBER HIM FOR
THE RIGHT REASONS

"I find my greatest pleasure, and so my reward, in the work that precedes what the world calls success."

— THOMAS A. EDISON

❧❦☙

It is always interesting how history remembers specific facts and not others. Many facts can get muddled over a time where people get forgotten or falsely credited. Just like people thinking that Henry Ford invented the first automobile, or that Alexander Graham Bell invented the telephone or that Galileo invented the telescope.

❧❦☙

Thomas Alva Edison seems to be remembered these days as

either the man who invented the light bulb or the man who didn't. Without knowing any more about him, you are either giving him false praise for something he didn't do or not taking into account all the other work that he did.

❧

In truth, Edison was a man who invented a lot of things while bringing about incredible advancements in many other things. With the stock ticker, the telegraph, the light bulb, and motion picture, he may not have invented them, but he improved on them beyond any recognition. They wouldn't be the technologies they were today without Edison's great mind working on them.

❧

What he did invent was astounding too. The phonograph was a truly remarkable breakthrough, and he can claim to be the first person ever to hear recorded sound isn't something that should be underestimated. His other inventions helped change the world around him and bring about a modern age in technology.

❧

Edison is perhaps one of the most misunderstood minds over the last few centuries; it's interesting to see what his reputation has become today. He will forever be known by many as the inventor of the light bulb, but it would be even more impressive to be known as the first person have made a recorded sound, which he actually did.

❧

At Menlo Park, Edison also changed the way that he thinks about teamwork and invention. He showed that bringing great minds together can create great things. You look at Google's '*Googleplex*' facility in California, and you can't help but see it as a modern interpretation of what Edison achieved at Menlo Park.

❧

Edison is rightfully remembered as one of the greatest inventors of all time, but often for the wrong reasons. The next time someone mentions that Edison invented the light bulb, you can put them right and tell them what he did do. And the next time someone mentions that Edison shouldn't be given credit as he didn't invent the light bulb, you can also inform them of all the great things he should be given credit for.

❧

Edison gave the world light, and he gave the world sound. It is an incredible legacy for a man who's rather fled from Canada, was turned away from school for being addled, and had severe hearing difficulties. He achieved incredible things and did it all through hard work. A misunderstood man in many respects, but one of the greatest to have ever lived.

THE BEST BOOKS ON THOMAS EDISON

- Thomas Edison: Inventing the Modern World –
  Alexander Kennedy – A highly acclaimed book
  about how the man changed the world.
- Edison and the Rise of Innovation – Leonard De
  Graaf – A look at how Edison changed the way
  that we think about invention and innovation.
- The Wizard of Menlo Park: How Thomas Alva
  Edison Invented the Modern World – Randall E.
  Stross – A look at how the world that Edison was
  born in was so vastly different to the world he
  died in.

YOUR FREE EBOOK!

As a way of saying thank you for reading our book, we're offering you a free copy of the below eBook.

Happy Reading!

GO WWW.THEHISTORYHOUR.COM/CLEO/

Made in the USA
Las Vegas, NV
12 February 2024

85709193R00059